Jeremy Clarkson

The Life and Legacy

of an

English Television Presenter,

Journalist, Farmer,

and

Author

BY

Brown Emily

Copyright © 2025 Brown Emily

TABLE OF CONTENTS

INTRODUCTION

The name Jeremy Clarkson evokes strong feelings right away. He is regarded by some as the voice of British motoring, influencing the opinions of millions of people toward automobiles, speed, and adventure.

Others view him as a provocateur whose unreserved approach and frank humor have caused controversy for decades.

But beneath the headlines and public image lies a complex guy whose work includes farming, television, journalism, and writing; he is a man whose impact extends well beyond the automotive industry.

Clarkson was born in Doncaster, South Yorkshire, in 1960. Although his early years were uneventful, they laid the groundwork for a personality that would eventually become legendary.

His curiosity about the universe, along with his restless energy and propensity to defy expectations, was evident from an early age.

These characteristics would define his career, making him someone who actively transformed the media rather than just participating in it.

He stood out in a crowded field of journalists and broadcasters thanks to his unique voice, which was characterized by wit, irreverence, and bravery.

As a writer, Clarkson first became well-known for his essays that fused humor, insight, and astute observation.

His ability to engage readers via storytelling and honesty was evident in his early work and would later become a defining characteristic of his career.

However, his effect was heightened by television. Clarkson revolutionized automotive journalism as the face of Top Gear, fusing entertaining and knowledge to make a specialized topic a worldwide sensation.

His capacity to entertain, educate, and challenge thinking all at once was demonstrated by the show, which evolved into more than just a car-related program.

What is the significance of Jeremy Clarkson? His importance stems from his unique blend of skill, charisma, and fearlessness rather than just his accolades or audience size.

In a time when public figures are frequently circumspect, Clarkson has continuously defied expectations, whether by challenging the current quo, participating in political and cultural discussions, or voicing viewpoints that others would find offensive.

He shows that both entertainment and journalism can be fearless, captivating, and open to new ideas.

The impact of Clarkson goes beyond television. He discusses his thoughts on vehicles, life, and society with his trademark candor and humor as a writer and author.

His life experiences, which range from farming endeavours to travel and adventure, provide us with a glimpse into the guy behind the public image and serve as a reminder that beneath the news is a person who is driven by ambition, curiosity, and a love of life.

Recognizing that Jeremy Clarkson is more than media soundbites and controversy is where this book starts. His life and work serve as an example of creativity, resiliency, and the strength of voice.

We can appreciate his influence on journalism and television, as well as the lasting legacy he continues to leave in media and entertainment, by getting to know the man behind the headlines.

This is the tale of an unabashed existence, a profession that defies simple classification, and a character that has influenced modern culture for better or ill.

CHAPTER ONE

Early Life and Formative Years

On April 11, 1960, Jeremy Clarkson was born in Doncaster, South Yorkshire, into a family that embodied both traditional British values and ambition.

His mother, Shirley Gabrielle Clarkson, created a nurturing and orderly environment, while his father, Edward Grenville Clarkson, was a former pilot in the Royal Air Force who went on to become a sales manager.

Jeremy was raised in a middle-class home where he was encouraged to pursue his interests while also realizing the importance of education and life skills, a balance that would shape much of his future.

Clarkson displayed a combination of curiosity and mischievousness from a young age. His early years were characterized by animated family debates and a sense of humor that would later define his public presence.

He showed an early interest in mechanics by enjoying disassembling objects, playing with bicycles, and figuring out how machines operated.

His subsequent career in journalism and television would be shaped by the practical approach and inquisitive mindset he gained from these encounters. A major influence on Clarkson's character development was his education.

Teachers at the Don caster primary school acknowledged his keen intelligence, witty humor, and occasional rebellious spirit. Later, he went to Repton School in Derbyshire, where his humor and independent thought continued to make him stand out.

He thrived in fields that allowed for creativity and self-expression, even though traditional academics were occasionally difficult for him.

He became interested in English writing and literature, contributing to school publications and demonstrating an early talent for telling gripping stories.

His success as a writer and broadcaster would subsequently be largely attributed to these abilities. Clarkson's initial forays into journalism and the media were tiny but significant.

He tried writing articles for minor journals and local newspapers when he was a youngster. These initial efforts demonstrated a gift for observation and a skill for transforming ordinary events into captivating tales.

He was naturally able to blend intelligence and humor, a talent that would come to define his later writing.

In addition, his fascination with radio and television introduced him to the wider realm of media and broadcasting, setting the stage for a profession that would combine his loves of communication, vehicles, and storytelling.

These early years helped mould the man who would go on to become one of Britain's most well-known media figures. They were influenced by a loving family, a solid school setting, and early media experimentation.

During this time, he gained curiosity, bravery, and storytelling skills that prepared him for a profession characterized by humor, sharp opinions, and a willingness to confront conventions.

Learning information and skills was only one aspect of Jeremy Clarkson's early years; he also had to find his voice, gain confidence, and start a lifetime interest in the media.

CHAPTER TWO

Breaking into Journalism

Curiosity and a desire to convey his observations in a humorous and truthful manner drove Jeremy Clarkson to pursue a career in journalism.

Prior to being well-known on television, Clarkson experimented with writing as a means of understanding vehicles, daily life, and the peculiarities of British society.

Despite being modest, his early work was essential in establishing the style that would come to identify him.

He began by writing brief freelancing articles for regional newspapers and magazines, where he gained experience in entertaining and persuading readers.

He learned from these early encounters how crucial it is to capture the audience's interest right away and maintain it through the opening line.

Clarkson's voice became distinct from other writers as he grew more self-assured. He blended personal tales with criticism on more general cultural concerns, combining irreverence with astute observation.

He was able to be both entertaining and educational, whether he was discussing social trends or evaluating a car's performance.

His distinctive blend of humor and authority became his signature, enabling him to play with perspective and tone in ways that few other journalists dared.

For Clarkson, writing was a method to tell stories in his own distinctive style, not just a way to convey facts.

There was some controversy around Clarkson's early career as well. His unvarnished thoughts occasionally unnerved editors and readers, even in his first published piece.

His willingness to question conventions became a defining characteristic, even though some people thought his approach was controversial.

Clarkson attracted notice by being honest and outspoken, in contrast to those who adhered to predetermined formulas.

His advancement was not hampered by these early debates; rather, they demonstrated his will to stay true to himself and be open to receiving as much praise as criticism.

He gained prominence as his style developed. He was invited to write for bigger audiences by publications that started to provide more substantial opportunities.

His features and editorials were notable for their blend of humor, technical expertise, and unreserved opinion. Professionals in the field took notice, seeing in Clarkson a new voice that could infuse old subjects with new vitality.

His eventual move to television, where his storytelling prowess would reach an even larger audience, was made possible by this expanding reputation.

In retrospect, Clarkson's formative journalism years were a period of development and exploration. He honed his tone, experimented with limits, and discovered the importance of being truthful with readers.

He made a definite commitment to involvement and sincerity, which is why he garnered the attention and controversies he did.

These early encounters shaped Clarkson's identity as a journalist and storyteller and laid the groundwork for a career that would have a significant impact on popular culture and the media for years to come.

CHAPTER THREE

Television Breakthrough

Jeremy Clarkson's journey to television celebrity is directly related to his work on Top Gear, the BBC driving show that became both a cultural phenomenon and the defining platform of his career.

While Clarkson had already established notoriety as a columnist and journalist with a keen, often confrontational voice, television allowed him to reach a bigger audience and completely display his unique personality.

Joining Top Gear in the late 1980s, Clarkson brought a blend of wit, irreverence, and a real love of vehicles that immediately differentiated him from his contemporaries.

His approach to motorsport journalism was never restricted to technical reviews; he aimed to entertain, inspire thinking, and sometimes cause controversy, ensuring that every episode showed his unmistakable personality.

Clarkson's career on Top Gear generated countless memorable events that captivated viewers' attention and often made news.

From crazy tasks, like racing across countries in odd vehicles, to evaluating automobiles in harsh weather, his segments combined humor, spectacle, and true automotive intelligence.

Whether pushing a supercar to its limits on a track or embarking on epic road excursions with his co-presenters, Clarkson made regular test drives into storytelling events.

His ability to build suspense, humor, and anticipation boosted the show from a specialist automotive program to a mainstream entertainment phenomenon, attracting listeners who might not even have been car fans.

Alongside co-presenters Richard Hammond and James May, Clarkson cultivated a dynamic that became a hallmark of the show, merging competitive camaraderie, banter, and moments of unexpected vulnerability that made the hosts approachable and entertaining.

Beyond his unique exploits and standout performances, Clarkson changed the tone and aesthetic of British television.

His audacious remarks, readiness to question norms, and sporadic arousal of controversy demonstrated a move toward personality-driven programming.

Top Gear under Clarkson redefined what factual entertainment might be by focusing on entertainment, spectacle, and pushing boundaries in addition to vehicles. The combination of humor, courage, and experience was well received by British viewers.

The program proved that viewers were interested in the personality of the presenter as much as the facts, and it became a model for how factual content could be both entertaining and profitable.

The legacy that followed was made possible by Clarkson's breakthrough on Top Gear. It demonstrated that a journalist may become well-known by showcasing his ability to blend expertise, comedy, and acting into engaging television.

By demonstrating that factual programming could be engaging and entertaining without compromising content, his work helped revolutionize British television.

Over thirty years later, his impact, style, and the iconic moments he produced are still relevant, confirming his status as one of the most significant people in contemporary television.

Jeremy Clarkson became a cultural hero because to Top Gear, which was more than simply a show.

CHAPTER FOUR

Expanding Horizons

Jeremy Clarkson could not settle for a single position after becoming a television celebrity with Top Gear.

He was always looking for new ways to demonstrate his versatility, taking on projects that brought out various facets of his character and abilities.

Although Clarkson is most recognized for his audacious appearances on auto television, his work outside of Top Gear demonstrates a professional willing to interact with viewers in a variety of ways.

After Top Gear, Clarkson's television endeavours showed that he was not afraid to experiment with other formats and take chances.

A major break from motoring was made with Clarkson's Farm, which offered an open and frequently amusing account of his experiences managing a farm in the English countryside.

The show gives fans a closer look at Clarkson by capturing his struggles, victories, and tenacity.

Furthermore, The Grand Tour gave him the opportunity to carry on with his automotive exploits on a worldwide scale, fusing entertainment, vehicles, and travel in a novel, cinematic manner.

These performances demonstrated that Clarkson could maintain his individuality while modifying his manner for a range of settings.

Clarkson ventured into radio, podcasts, and live engagements in addition to television. He was able to address audiences directly on these channels in a more casual and unvarnished manner.

He was able to humorously and perceptively convey his thoughts on current affairs, motorsport, and personal experiences, especially through podcasts.

His ability to engage live audiences with spontaneity and wit that translate differently from the screen was demonstrated during public appearances at charity events and industry gatherings.

Another important aspect of Clarkson's work has been writing. His diverse interests and distinctive voice are evident in his books and columns.

Clarkson's writing blends humor and insight, making it both amusing and educational, whether he is making observations on motoring, rural life, or more general cultural issues.

While his books provide a more in-depth examination of his experiences and viewpoints, his writings for newspapers and publications appeal to readers in the same captivating, opinionated manner as his television appearances.

Jeremy Clarkson's career has grown significantly beyond automotive television with live events, podcasts, radio, television, and writing.

These initiatives demonstrate a professional who is open to trying new things while maintaining his unique flair. His impact now extends across other media, establishing a varied, captivating, and long-lasting legacy.

CHAPTER FIVE

Personality and Public Perception

Jeremy Clarkson is renowned for both his accomplishments in his career and his personality. His exaggerated character has been a major source of criticism as well as fuel for his success.

Since the beginning of his career, Clarkson has gained millions of admirers and generated controversy with his unvarnished honesty and unreserved candor.

His remarks, whether made on TV, in print, or on social media, have regularly sparked discussions and occasionally touched on subjects that many people view as delicate or politically sensitive.

These disputes, which have periodically sparked public outrage and official censure, have ranged from scathing criticisms of well-known individuals to thought-provoking remarks on societal concerns.

However, far than making him less famous, these episodes have strengthened the idea that Clarkson is a free-thinking individual, which both alienates and draws people.

Clarkson's public presence revolves around humor, which is intimately related to his candor. His humorous approach frequently blends self-deprecation, exaggeration, and satire to produce amusing and incisively perceptive moments.

Clarkson has a gift for transforming even commonplace topics into captivating entertainment, whether he is making commentary on vehicles, societal trends, or current affairs.

His humor relies on boldness and surprise and is rarely careful or politically acceptable. Although this strategy has occasionally been criticized for being insensitive, it has also added to his ongoing popularity, especially with audiences that respect irreverence and sincerity.

Particularly in the field of automotive journalism, Clarkson's ability to blend humor and authority has created a unique voice that is instantly identifiable.

Clarkson has a complicated and illuminating relationship with both his fans and the media. On the one hand, those who respect his candor, charm, and distinct style are fiercely loyal to him.

Support is frequently shown by fans on social media or in public defense of a cause during a dispute. In a media environment that is sometimes characterized by caution, he is widely regarded as an advocate of honesty and personal freedom.

However, his public character exposes him to ongoing media attention. His remarks and deeds are regularly scrutinized, magnified, and occasionally sensationalized.

This has led to a tug-of-war relationship between him and the media: he gains from publicity but frequently objects to the way his remarks are portrayed.

His public life now revolves around striking this balance, which emphasizes both the benefits and drawbacks of celebrity.

Ultimately, Jeremy Clarkson's work and public reputation are inextricably linked. His reputation and the impact he has had on journalism and television are defined by his humor, outspokenness, controversy, and interactions with fans and the media.

His public life serves as an example of how a powerful, divisive personality can influence public opinion, captivate audiences, and have a long-lasting effect on popular culture.

CHAPTER SIX

Life Beyond the Camera

Although Jeremy Clarkson is well-known for his audacious television image, passion for automobiles, and unvarnished thoughts, he has an another side that is hidden from the public eye.

Clarkson has developed a life centred on the countryside, hobbies, and family ties away from cameras and media attention, revealing a quieter, introspective, and grounded side of himself.

Beyond television, farming and rural living play a big role in Clarkson's life. He took to the routines and difficulties of farming after buying a farm in the Cotswold's.

Clarkson addresses the practical skills, patience, and problem-solving required to manage cattle, tend crops, and navigate the uncertainties of farm work with excitement.

His farm serves as more than simply a setting for comedies or TV shows; it is a real, dynamic business that has taught him valuable lessons about perseverance, diligence, and the joy of real accomplishments.

Clarkson is able to interact directly and meaningfully with nature because of the stark contrast between the unpredictable nature of farm life and the regulated setting of television.

Clarkson engages in a variety of pastimes and interests outside of farming, which reflects his natural curiosity and practical outlook on life.

Even though vehicles are still his main hobby, he also likes to garden, experiment in the cooking, and look at vintage equipment.

Outside of work, these activities offer a kind of intellectual and creative stimulation. Clarkson enjoys doing things that let him think, tinker, and solve problems on his own terms, whether that be fixing cars, picking up new skills, or just travelling the countryside.

His pastimes serve as a reminder that his passions are based on personal fulfillment and exploration rather than celebrity and media attention.

Family life is just as significant in Clarkson's personal life. He keeps tight ties with his kids and other family members in spite of the public exposure he gets.

He takes a humorous, sincere, and considerate approach to family life, cherishing private times of bonding.

Basic hobbies, weekend schedules, and time spent with loved ones are essential to his wellbeing because they give him a sense of balance and stability that comes with not being famous.

This intimate aspect reveals a man who is devoted to his loved ones, exhibiting affection, dedication, and a desire to make enduring memories.

Essentially, aspects of Jeremy Clarkson's personality that diverge from his public persona are revealed by his life away from the camera.

He interacts with nature firsthand and learns about the benefits of perseverance through farming.

His interests provide him with stimulation and relaxation while enabling him to experiment, create, and learn.

Additionally, his family life places a strong emphasis on stability, love, and connection.

When taken as a whole, these features present Clarkson as more than just a TV personality; they depict a guy who finds identity, happiness, and purpose in life away from the limelight.

CHAPTER SEVEN

Contributions to Journalism and Entertainment

Jeremy Clarkson's impact on entertainment and media extends well beyond his audacious television character.

Although he is best known for being the controversial host of Top Gear and The Grand Tour, he has made substantial and enduring contributions to automotive journalism and the media in general.

By fusing knowledge, humor, and strong opinions in ways that were unusual prior to his emergence, Clarkson transformed the way that car culture is presented.

In addition to covering automobiles, he made automotive journalism approachable, amusing, and frequently thought-provoking, making it appealing to both vehicle fanatics and general audiences.

He established a new benchmark for audience engagement by demonstrating via his work that journalism could be both entertaining and educational.

Clarkson's unique approach is essential to his influence. Viewers were able to relate to him as a personality rather than merely a broadcaster because of his unabashedly hilarious, opinionated, and intimate voice.

In order to make complicated automobile facts understandable and entertaining, his assessments frequently blended technical analysis with tales, humor, and the occasional exaggeration.

This method deviated from conventional, formal auto evaluations and inspired other reporters to use personality, storytelling, and inventiveness in their writing.

A generation of authors, presenters, and content producers were encouraged to experiment with tone and format by Clarkson's demonstration that passion and knowledge could coexist with amusement.

Beyond his own sense of style, Clarkson has a wide-ranging impact on the media.

By demonstrating that automobiles could be at the heart of gripping media and television stories, he contributed to the mainstreaming of automotive programming.

Programs like Top Gear attracted millions of viewers worldwide and shaped the way other shows approach infotainment, making them cultural phenomena.

By emphasizing spectacle, storytelling, and storylines driven by personality, Clarkson showed that journalists may be entertaining without compromising their trustworthiness.

His work further solidified his position as a figure who influenced media standards and expectations by igniting discussions about journalism ethics, style, and substance.

Collaboration and mentoring have also played significant roles in Clarkson's career. He collaborated closely with authors, producers, co-hosts, and crew members to foster creative situations.

He demonstrated the need of teamwork in creating captivating and memorable material by working with colleagues like James May and Richard Hammond.

His writing and public commentary have impacted younger journalists outside of television by demonstrating how to blend humor, knowledge, and personality in the media.

Even though his strategy has occasionally generated controversy, it has also inspired up-and-coming journalists and artists to look for creative methods to engage audiences.

In conclusion, Jeremy Clarkson has made significant and influential contributions to both entertainment and journalism.

With his distinct style, he transformed automotive journalism, impacted how media integrates enjoyment and information, and promoted cooperation and mentoring among industry professionals.

In addition to his shows and writings, he established new guidelines and methods that have inspired others to approach journalism and entertainment with originality, individuality, and self-assurance.

Clarkson's work has a lasting impact on the media environment and the automotive industry, proving that journalism can simultaneously educate, entertain, and inspire.

CHAPTER EIGHT

Lessons from a Bold Career

Jeremy Clarkson's career serves as an excellent illustration of how tenacity, enthusiasm, and a powerful public image may influence achievement.

Clarkson's journey from his early days as a journalist to becoming one of the most well-known television personalities in the world demonstrates that making a lasting impression takes work, taking chances, and being ready to stand out.

The value of perseverance is a key takeaway from his career. It's never easy to break into the journalism and television industries, but Clarkson gradually made his mark.

He persistently sought chances that aligned with his interests and talents despite rejection, skepticism, and fierce competition.

For Clarkson, perseverance meant more than simply remaining in the business; it meant creating a distinctively his voice.

His dedication to perfecting his art, whether through newspaper columns or television presentations, demonstrates how perseverance and self-assurance may result in noteworthy achievements.

His achievement has also been characterized by his passion. The energy underlying Clarkson's work is fueled by his passion for vehicles, the countryside, and narrative.

His writing, public appearances, and side ventures all reflect this sincere interest, as does his on-screen persona. Passion makes it possible for work to engage listeners more deeply.

As demonstrated by Clarkson's career, work that is motivated by genuine interest rather than duty strikes a deeper chord and makes a greater impression.

Clarkson's career has been marred by controversy and criticism, thus his handling of public scrutiny is a valuable lesson in and of itself.

His candidness has frequently generated controversy and strong feelings. His experience serves as a reminder that public boldness carries responsibilities.

Anyone with a prominent career needs to learn how to handle criticism while remaining genuine.

Clarkson demonstrates that, if one knows how to strike a balance between accountability and honesty, it is possible to maintain credibility and keep having an impact in a profession despite coming under scrutiny.

His life and work reveal several important lessons. First, individualism counts: in highly competitive fields, taking chances and publicly voicing one's thoughts might make one stand out. Second, resilience is crucial.

Although obstacles and failures are unavoidable, overcoming them improves one's character and professional acumen.

Third, passion improves performance: when sincere excitement drives work, it increases involvement and creates a lasting imprint.

Last but not least, maintaining success requires striking a balance between one's personal principles and public impression.

Clarkson's journey demonstrates that when one stays true to themselves while acknowledging the wider effects of their acts, boldness and efficacy may coexist. In conclusion, Jeremy Clarkson's career serves as a lesson in tenacity, drive, and public image management.

His accomplishments and conflicts teach journalists and media professionals as well as anyone else who wants to follow their career with passion and uniqueness.

Learning from his experience can help one develop a unique career while appreciating the value of fortitude, self-awareness, and the guts to adopt a fearless professional identity.

CONCULSION

The career of Jeremy Clarkson is one of those exceptional tales that defies easy classification. He is simultaneously a farmer, novelist, journalist, and television host, a man whose impact transcends all of his vocations.

In retrospect, it is evident that his journey has involved both entertaining audiences and influencing public opinion.

Boldness, wit, and an uncompromising dedication to doing things his own way have characterized Clarkson's career from his early days in journalism to his legendary time on Top Gear and beyond.

It takes more than a list of accomplishments to sum up such a career; it also entails acknowledging the unique voice he brought to each platform he touched.

Clarkson left behind a complex legacy. He revolutionized how viewers interact with automotive media on television.

His approach, which combined irreverence, humor, and unvarnished honesty, made programs like Top Gear a worldwide sensation.

He wasn't merely showcasing automobiles; he was crafting memorable moments that viewers would think about long after the engines had subsided.

In addition to providing amusement, he stimulated debates with his frank views on culture, society, and the nature of stardom.

Even though his strategy occasionally attracted criticism, it also cemented his standing as a distinctive and significant character in British journalism.

In addition to the ratings he received, his influence is evident in the way a new generation of journalists, authors, and broadcasters feel free to use personality and provocations in their work.

Clarkson's influence can also be seen in his writing, which includes magazine articles, books, and newspaper columns. He entertains while promoting introspection by fusing humor, wisdom, and opinion.

This dichotomy was thoughtful and hilarious, confrontational yet genuine encapsulates his career: a man who isn't scared to take chances, even if they lead to controversy.

His art offers lessons that go beyond amusement, highlighting the importance of authenticity and the potency of a unique voice.

In the future, Clarkson's impact will only grow. He has demonstrated how a career based on character, moral character, and curiosity can change with the times.

He continues to defy expectations, reminding audiences and colleagues that reinvention is always possible, whether he is testing the newest cars, discussing society, or examining life on his farm.

In the end, reflecting on the past reveals a career that is both unorthodox and significant. In addition to being a list of accomplishments, Jeremy Clarkson's tale is a tribute to uniqueness, tenacity, and the guts to voice one's opinions.

His legacy is found in the amusement he has produced, the discussions he has sparked, and the expectations he has set for his successors.

His impact will endure throughout time, not only in journalism and television but also in the larger cultural environment he has influenced.

Clarkson's story shows that a profession based on integrity, wit, and courage can have a profound effect on society.

Printed in Dunstable, United Kingdom

72950469R00030